What We Know about Climate Change

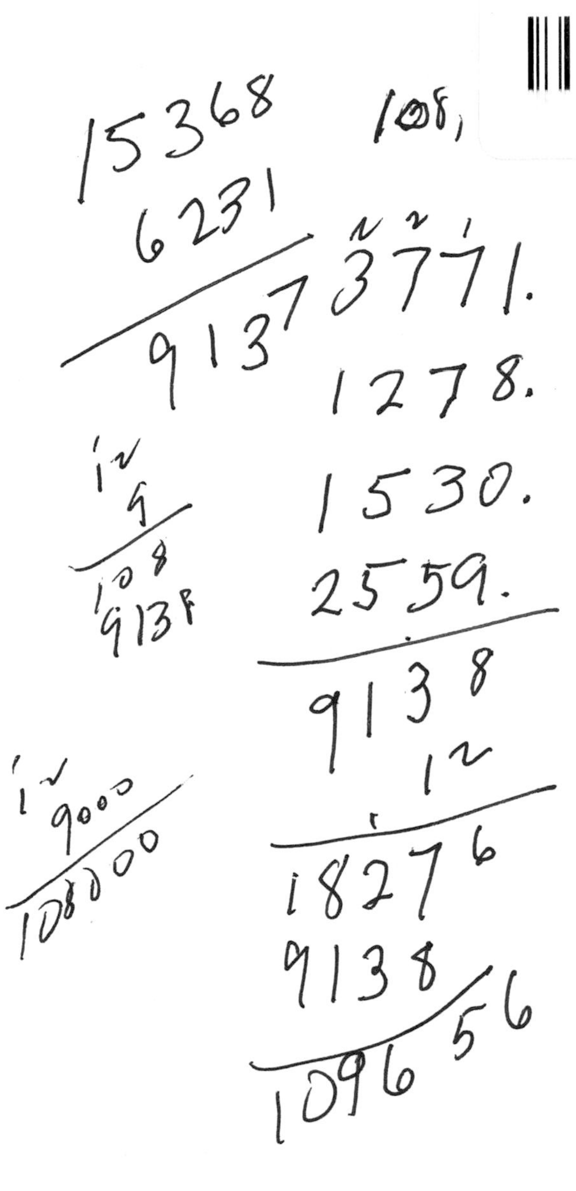

What We Know about Climate Change

Updated Edition

Kerry Emanuel
Foreword by Bob Inglis

The MIT Press
Cambridge, Massachusetts
London, England

This book was set in ITC Stone by Jen Jackowitz. Printed and bound in the United States of America.

Library of Congress Cataloging-in-Publication Data

Names: Emanuel, Kerry A., 1955- author.
Title: What we know about climate change / Kerry Emanuel ; with a new foreword by Bob Inglis.
Description: Updated edition [2018 edition]. | Cambridge, MA : The MIT Press, [2018] | Includes bibliographical references and index.
Identifiers: LCCN 2018010196 | ISBN 9780262535915 (pbk. : alk. paper)
Subjects: LCSH: Climatic changes. | Greenhouse effect, Atmospheric. | Global warming.
Classification: LCC QC903 .E43 2018 | DDC 363.738/74--dc23 LC record available at https://lccn.loc.gov/2018010196

10 9 8 7 6 5 4 3 2

Contents

Foreword

Bob Inglis

I once got in trouble with a friend, a person who shares my faith, when I told him that we should *celebrate* the science of climate change.

"Hold on," he objected. "I don't know about 'celebrating' the science. That's going too far."

I'm going to send my friend a copy of this primer by Kerry Emanuel. If he reads it, surely my friend will come to celebrate the science.

In this beautifully and accessibly written work, Emanuel takes us on a stunning review of the science. It's marvelous to get a glimpse of what we've come to know through the brilliant work of scientists like Emanuel. And yet, he freely admits that "We are humbled by a sense of ignorance." There's more to know than we'll ever know, and uncertainty will always drive toward new discoveries.

Of his fellow scientists, Emanuel tells us, "We are, most of us, driven by a passion to understand nature, and that compels us to be dispassionate about pet ideas. Partisanship—whatever its source—is likely to be detected by our colleagues and hurt our credibility, the true stock of the trade."

Of us—the public—Emanuel gets it exactly right, "As long as we continue to elect and appoint people who exaggerate what we don't know at the expense of what we do, intelligent debate will elude us. Time will be lost to inane talk of hoaxes and conspiracies, while other nations, by developing and implementing advanced forms of energy production, will gain an important economic advantage over the United States."

The risks of climate change carry with them the offer of nobility for the generation that rises to the challenge. As Emanuel points out, "[T]here are few, if any, historical examples of civilizations consciously making sacrifices on behalf of descendants two or more generations removed." We have the opportunity to be the stunning exception to that rule.

US Representative (R-SC4, 1993–1999 and 2005–2011)
Executive Director, republicEn.org

Preface

Since *What We Know about Climate Change* was first written more than a decade ago, climate science has continued to make important and far-reaching advances, while the signals of climate change have become ever more prominent and worrisome. Sixteen of the seventeen warmest years in the 136-year record of global mean surface temperature have occurred since 2001, and 2016 was the hottest year on record. Carbon dioxide continues its seemingly inexorable climb and now exceeds 400 parts per million, its highest level in millions of years. Sea level continues to rise and, absent serious measures to curtail greenhouse gas emissions, is projected to increase by another 1–3 feet by the end of this century, putting many coastal habitations in jeopardy. Typhoon Haiyan of 2013 set the world record for highest wind speed in a tropical cyclone, a record that was shattered just two years later by eastern Pacific Hurricane Patricia. Some of the more spectacular heat waves, wildfires, droughts, and floods of the last decade have been formally attributed to anthropogenic climate change. The world's foremost defenders of national security, such as the US Department of Defense, have come to recognize climate change as a significant security threat, as water and food shortages drive increasing immigration pressures that cause or exacerbate armed conflict.

Yet there are signs that, slowly, the world is preparing to confront the climate problem. A recent global poll conducted by the Pew Research Center places climate change as a leading global threat, second only to ISIS. As of 2017, 195 nations had signed on to the Paris Agreement, a serious attempt to hold the increase in global temperature since pre-industrial times to below 2°C.

Alas, for every two steps we take to address climate change, we insist on one step backward. This year, the United States withdrew from the Paris Agreement, thereby ceding international leadership and a potential global market of $6 trillion in carbon-free energy to China, currently the leading producer of renewable energy technology and a leader in the development of other carbon-free energy sources such as nuclear power. And Germany used its near-miraculous increase in renewable energy to replace carbon-free nuclear power rather than carbon-emitting coal, trumping a concern for future climate change with a fear of nuclear energy.

This third edition of *What We Know about Climate Change* updates the second edition by documenting climate change that has occurred since the second edition was published in 2012, and by updating the state of climate science and climate politics. As with the two previous editions, the intent is to provide a broad, readable overview of climate science rather than a compendium of evidence; in a new further reading section, I point the reader toward more extensive treatments of the subject.

What We Know about Climate Change

The Myth of Natural Stability

Two strands of environmental philosophy run through the course of human history. The first holds that the natural state of the universe is one of infinite stability, with an unchanging earth anchoring the predictable revolutions of the sun, moon, and stars. Every scientific revolution that challenged this notion, from Copernicus's heliocentricity to Hubble's expanding universe, from Wegener's continental drift to Heisenberg's uncertainty and Lorenz's macroscopic chaos, met with fierce resistance from religious, political, and even scientific hegemonies.

The second strand also sees stability as the natural state of the universe but holds that human beings destabilize it. The great floods described in many religious traditions are portrayed as attempts by a god or gods to cleanse the earth of human corruption. Deviations from cosmic order, such as meteors and comets, were more often viewed as omens than as natural phenomena. In Greek mythology, the blistering heat of Africa and the burnt skin of its inhabitants were attributed to Phaëthon, an offspring of the sun god Helios. Having lost a wager to his son, Helios was obliged to allow him to drive the sun chariot across the sky. In this primal environmental catastrophe, Phaëthon lost control and scorched the earth, killing himself in the process.

These two fundamental ideas—cosmic stability and man-made disorder—have permeated many cultures through much of history. They strongly influence views of climate change even today.

In 1837 Louis Agassiz provoked public outcry and scholarly ridicule when he proposed that many enigmas of the geologic record, such as peculiar scratch marks on rocks, and boulders far removed from their bedrock sources, could be explained by the advance and retreat of huge sheets of ice. His proposal marked the beginning of a remarkable endeavor, today known as paleoclimatology.

Paleoclimatology uses physical and chemical evidence from the geological record to deduce changes in the earth's climate over time. This field has produced among the most profound yet least celebrated scientific advances of our era. We now have exquisitely detailed knowledge of how climate has varied over the last few million years and, with progressively less detail and certainty, how it has changed going back to the age of the oldest rocks on our 4.5-billion-year-old planet.

For those who take comfort in stability, there is little consolation in this history. In just the past three million years, our climate has swung between mild states—similar to today's and lasting 10,000 to 20,000 years—and periods of 80,000 years or so in which giant ice sheets, in some places several miles thick, covered northern continents. Even more unsettling is the suddenness with which the climate can change, especially as it recovers from glacial periods.

Over longer intervals of time, the climate has changed even more radically. During the early part of the Eocene era, around 50 million years ago, the earth was free of ice, and giant trees grew on islands near the North Pole, where the annual mean temperature was about 60°F, far warmer than today's mean of about 30°F. There is also some evidence that the earth was almost entirely covered with ice at various times around 500 million years ago. These "snowball earths" alternated with exceptionally hot climates.

What explains these changes? For climate scientists, ice cores in Greenland and Antarctica provide intriguing clues about the great glacial cycles of the past three million years. As the ice formed it trapped bubbles of atmosphere, whose chemical composition—including, for example, its carbon dioxide and methane content—can now be analyzed. Moreover, it turns out that the ratio of two isotopes of oxygen locked up in the molecules of ice is a good indicator of the air temperature when and where the ice was formed. And the age of the ice can be determined by counting the layers that mark the seasonal cycle of snowfall and melting.

Relying on such analyses of ice cores and similar analyses of sediment cores from the deep ocean, researchers have learned something remarkable: the ice-age cycles of the past three million years were almost certainly caused by periodic oscillations of the earth's rotation and orbit that affect primarily the orientation of its axis. These oscillations do not much affect the *amount* of sunlight that reaches the earth, but they do change the *distribution* of sunlight with latitude. Ice ages occur when, as a result of orbital variations, arctic regions intercept relatively little summer sunlight so that ice and snow do not melt as much as they otherwise would.

The timing of the ice ages, then, is the result of the earth's orbit. It is discomfiting that these large climate swings—from glacial to interglacial and back—are caused by relatively small changes in the distribution of sunlight with latitude. Thus, on the time scale of ice ages, climate seems exquisitely sensitive to small perturbations in the distribution of sunlight.

And yet for all this sensitivity, the earth never suffered a permanent catastrophe of fire or ice. In the fire scenario, the most effective greenhouse gas—water vapor—accumulates in the atmosphere as it warms. The warmer the atmosphere, the more water vapor it can contain; as more water vapor accumulates, more heat gets trapped, and the warming spirals upward. This feedback, unchecked, is called the runaway greenhouse effect, and it continues until the oceans have all evaporated, by which time the planet is unbearably hot. One has to look only as far as Venus to see the end result. Any oceans that may have existed on that planet evaporated eons ago, yielding a super greenhouse inferno and an average surface temperature of about 900°F.

Death by ice can result from another runaway feedback. As snow and ice accumulate progressively equatorward, they reflect an increasing amount of sunlight back to space, further cooling the planet until it freezes into a snowball earth. As discussed above, there is some evidence that this actually happened to the earth several times around 500 million years ago. It used to be supposed that once the planet reached such a frozen state, reflecting almost all sunlight back to space, it could never recover. More recently it has been theorized that without liquid oceans to absorb the carbon dioxide continuously emitted by volcanoes, the gas would accumulate in the atmosphere until its greenhouse effect was finally strong enough to start melting the ice. Once begun, the positive feedback between temperature and reflectivity would work in reverse, rapidly melting the ice and leading to a hothouse climate in a short time. It would not take much change in the amount of sunlight reaching the earth to cause a snowball or runaway greenhouse catastrophe.

But solar physics informs us that the sun was about 25 percent dimmer early in the earth's history, which should have led to an ice-covered planet, a circumstance not supported by geological evidence from that era. So what saved the earth from an ice catastrophe?

Life itself, perhaps. Our atmosphere is thought to have originated in gases emitted from volcanoes, but the composition of volcanic gases bears little resemblance to air as we know it today. We believe that the early atmosphere consisted mostly of water vapor, carbon dioxide, sulfur dioxide, chlorine, and nitrogen. There is little evidence of much oxygen before the advent of cyanobacteria, a phylum of bacteria that produced oxygen through photosynthesis and began the transformation of the atmosphere into something like today's, consisting mostly of nitrogen and oxygen with trace amounts of water vapor, carbon dioxide, and other gases. Carbon dioxide content probably decreased slowly over time owing to chemical weathering—chemical reactions involving rainwater and rocks—possibly aided by biological processes. As the composition of the atmosphere changed, the net greenhouse effect weakened, compensating for the slow but inexorable brightening of the sun.

This compensation may not have been an accident. In the 1960s James Lovelock proposed that life actually exerts a stabilizing influence on climate by producing feedbacks favorable to it. He called his idea the Gaia hypothesis, named after the Greek earth goddess. But even according to this view, life is preserved only in the broadest sense: individual species, such as those that transformed the early atmosphere, altered the environment at their peril.

Clearly life has profoundly altered our climate. We humans are merely the most recent species to do so.

Greenhouse Physics

As the last chapter's sketch of the planet's early climatic history shows, the greenhouse effect plays a critical role in the earth's climate, and no sensible discussion of climate could proceed without grasping its nature.

The greenhouse effect has to do with radiation—in this context meaning energy carried by electromagnetic waves—which include such phenomena as visible light, radio waves, and infrared radiation. All matter with a temperature above absolute zero emits radiation. The hotter the substance, the more radiation it emits and the shorter the average wavelength of that radiation. A fairly narrow range of wavelengths constitutes visible light. The average surface temperature of the sun is about 10,000°F, and the sun emits much of its radiation as visible light, with an average wavelength of about half a micron. (A micron is one millionth of a meter; there are 25,400 microns in an inch.) The earth's atmosphere emits radiation as though its average temperature were around 0°F, at an average wavelength of about 15 microns. Our eyes cannot detect this infrared radiation. It is important to recognize that the same object can both emit and absorb radiation: when an object emits radiation, it loses energy, cooling it; absorption, on the other hand, heats it.

Most solids and liquids absorb much of the radiation they intercept, and they also emit radiation rather easily. Air is another matter. It is composed almost entirely of oxygen and nitrogen, each in the form of two identical atoms bonded together in a single molecule. Such molecules barely interact with radiation: they allow free passage to both solar radiation moving downward to the earth and infrared radiation moving upward from the earth's surface.

If that were all there is to the atmosphere, it would be a simple matter to calculate the average temperature of the earth's surface: it would have to be just warm enough to emit enough infrared radiation to balance the shortwave radiation it absorbed from the sun. (Were it too cool, it would emit less radiation than it absorbed and would heat up; conversely, were it too warm it would cool down.) Accounting for the sunlight reflected back to space by the planet, this works out to be about 0°F, far cooler than the observed mean surface temperature of about 60°F.

Fortunately for us, our atmosphere contains trace amounts of other substances that do interact strongly with radiation. Foremost among these is water, H_2O, consisting of two atoms of hydrogen bonded to a single atom of oxygen. Because of its more complex geometry, it absorbs and emits radiation far more efficiently than molecular nitrogen and oxygen. In the atmosphere water exists both in its gas phase (water vapor) and its condensed phase (liquid water and ice) as clouds and precipitation.

Water vapor and clouds absorb sunlight and infrared radiation, and clouds also reflect sunlight back to space. The amount of water vapor in a sample of air varies greatly from place to place and time to time but does not usually exceed about three percent of the mass of the sample. Besides water, there are other gases that interact strongly with radiation, including carbon dioxide (CO_2, presently about 405 molecules for each million molecules of air) and methane (CH_4, around 1.9 molecules for each million molecules of air).

Collectively the greenhouse gases—water vapor, carbon dioxide, methane, and several others—are nearly transparent to sunlight. Were it not for the presence of clouds, short-wavelength radiation would pass virtually unimpeded to the surface, where most of it would be absorbed. On the other hand, these same gases absorb much of the long-wavelength, infrared radiation that passes through them. To compensate for the heating this absorption causes, the greenhouse gases must also emit radiation, and each layer of the atmosphere thus emits infrared radiation upward and downward.

As a result, the surface of the earth receives radiation from the atmosphere as well as from the sun. It is an extraordinary fact that, averaged over the planet, the surface receives almost twice as much radiation from the atmosphere as it does from the sun. To balance this extra input of radiation—the radiation emitted by atmospheric greenhouse gases and clouds—the earth's surface warms up and thereby emits more radiation itself. This is the essence of the greenhouse effect.[1]

If air were not in motion, the observed concentration of greenhouse gases and clouds would raise the average temperature of the earth's surface to around 85°F, much warmer than observed. In reality, hot air from near the surface rises upward and is continually replaced by cold air moving down from aloft. These convection currents lower the surface temperature to an average of 60°F while warming the upper reaches of the atmosphere. So the downward emission of radiation by greenhouse gases keeps the earth's surface warmer than it would otherwise be, and, at the same time, the convective movement of air dampens the warming effect and keeps the surface temperature bearable, while warming the upper atmosphere

The greenhouse gases collectively comprise roughly 0.3% of the mass of the atmosphere. Almost all of this is water vapor, which while highly variable in space and time, makes up on average about 0.25% of the atmosphere. But the concentration of water vapor is mostly just a function of temperature, and it adjusts to an equilibrium value in just a few weeks. Thus water vapor is properly considered a feedback in the climate system: warmer air has more water vapor, which through its greenhouse effect makes the system yet warmer.

By contrast, it takes hundreds to thousands of years for CO_2 concentrations to adjust naturally, so it is really long-lived greenhouse gases like CO_2 that exert a controlling influence on climate.

It is truly astonishing that, as the great Irish physicist John Tyndall discovered in the mid-nineteenth century, the tolerably warm conditions we enjoy are thanks to long-lived greenhouse gases like CO_2 that together constitute only about 0.04% of our atmosphere. On time scales of millions of years and greater, these long-lived greenhouse gases act as a kind of thermostat. For example, if the planet were to warm up appreciably, chemical weathering of rocks would accelerate, taking CO_2 out of the atmosphere, cooling the climate back toward it original equilibrium. Conversely, were the planet to cool, weathering would slow, allowing CO_2 to accumulate in the atmosphere thus warming it back to equilibrium.

In 1897, the Swedish chemist and Nobel laureate Svante Arrhenius realized that increasing combustion of fossil fuels would eventually raise CO_2 concentrations, simply because human emissions were far too large for the natural system to deal with on human time scales. By 1906 he had calculated that doubling the concentration of CO_2 would raise the earth's surface temperature by about 4°C, a number well within contemporary estimates of 2–4.5°C per doubling of CO_2. It is important to note that Arrhenius did this without the benefit of computers, relying on basic physics already fairly well quantified by his time. Figure 1 tests Arrhenius's prediction by comparing the observed natural logarithm of the CO_2 concentration with observed global mean temperature. While there are many natural influences on climate, such as small variations in solar output and

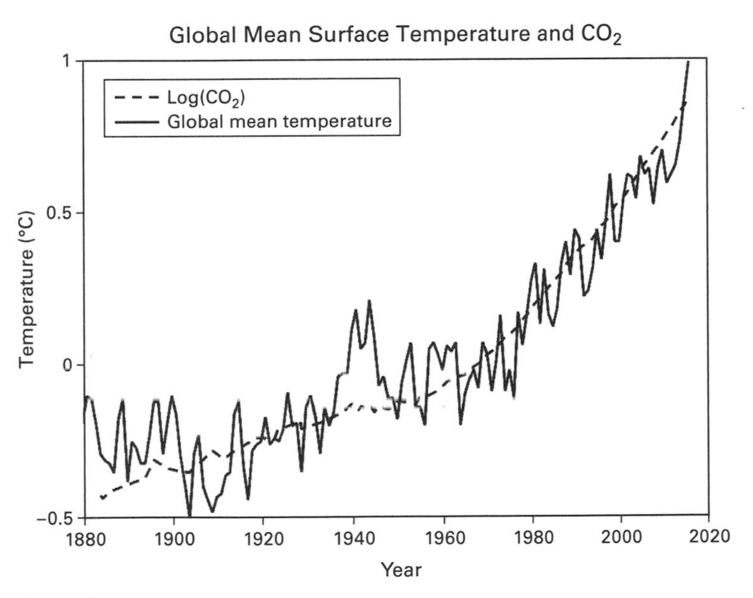

Figure 1
Natural logarithm of atmospheric CO_2 concentration (dashed), from ice cores and, after 1958, from direct measurements, compared to global mean temperature (solid) from the NASA Goddard Institute for Space Studies.

large volcanic eruptions, one can see that Arrhenius's prediction as so far been well verified.

The long residence time of CO_2 in the atmosphere implies that unless we can figure out an artificial way to remove it from the atmosphere, we will be stuck with increased levels of this important greenhouse gas and its associated climate anomalies for several millennia.

Why the Climate Problem Is Difficult

Basic climate physics is entirely uncontroversial among scientists.[1] And if one could change the concentration of a single greenhouse gas while keeping the rest of the system (except its temperature) fixed, it would be fairly simple to calculate the corresponding change in surface temperature. For example, doubling the concentration of CO_2 would raise the average surface temperature by about 1.9°F, enough to detect but probably not enough to cause serious problems.

But, of course, it's not actually that simple. Almost all of the uncertainty in climate science arises from the fact that, in reality, changing any single component of the climate system will indirectly cause other components of the system to change as well. These knock-on effects are known as feedbacks, and the most important and uncertain of these involves water.

There is a fundamental difference between water and most other greenhouse gases. Whereas a molecule of carbon dioxide or methane might remain in the atmosphere for hundreds or thousands of years, water is constantly recycled between the atmosphere, land surface, and oceans, so that a particular molecule of water resides in the atmosphere for, on average, about two weeks. On climate time scales, which are much longer than two weeks, atmospheric water vapor is tightly controlled by temperature and by physical processes operating within clouds. If one were to deposit a huge pulse of water vapor in the atmosphere, it would be gone in a few weeks.

Water vapor and clouds are the most important greenhouse substances in the atmosphere; clouds affect climate not only by emitting infrared radiation toward the surface and warming it up but also by reflecting sunlight back into space, thus cooling the planet.

Water is carried upward from its source at the surface by convection currents, which themselves result from greenhouse-induced warming of the surface. Simple physics as well as detailed calculations using computer models of clouds show that the amount of water vapor in the atmosphere is sensitive to the details of the physics by which tiny cloud droplets and ice crystals combine into larger raindrops and snowflakes, and how these in turn fall and partially re-evaporate on their way to the surface. The devil in these details seems to carry much authority with climate.

This complexity is limited, however, because the amount of water in the atmosphere is subject to a fundamental and important constraint. The concentration of water vapor in any sample of air has a strict upper limit that depends on its temperature and pressure. In particular, this limit rises very rapidly with temperature. The ratio of the actual amount of water vapor in a sample to this limiting amount is the familiar quantity called *relative humidity*. Calculations based on a large variety of computer models and observations of the atmosphere all show that as climate changes, relative humidity remains approximately constant. This means that as atmospheric temperature increases, the actual amount of water vapor increases as well. Water vapor is a greenhouse gas, though, and so increasing temperature increases water vapor, which leads to further increases in temperature. This positive feedback in the climate system is the main reason that the global mean surface temperature is expected to increase somewhat more than the 1.9°F that doubling CO_2 would produce in the absence of feedbacks. (At very high temperatures, the water vapor feedback can run away, leading to the catastrophe of a very hot planet with no oceans. This apparently happened on Venus, whose mean surface temperature is about 900°F, despite the fact that it absorbs less sunlight than earth, thanks to its very extensive and reflective cloud cover.)

The amount and distribution of water vapor in the atmosphere is also important in determining the distribution of clouds, which play a complex role in climate. On the one hand, they reflect about 22 percent of incoming solar radiation back to space, thereby cooling the planet. On the other hand, water vapor and clouds absorb solar radiation, and both absorb and emit infrared radiation, thus contributing to greenhouse warming. As anyone who has spent time gazing at the sky knows, clouds can assume beautiful and intricate patterns; capturing such patterns in computer models is challenging, to say the least. Thus it is hardly surprising that different global climate models produce different estimates of how clouds might change with changing climate. This is the largest source of uncertainty in climate-change projections.

A further complication in this already-complex picture comes from aerosols: minute solid or liquid particles suspended in the atmosphere. Industrial activity and biomass burning have brought about large increases in the aerosol content of the atmosphere, which most researchers agree have had a large effect on climate. Of the anthropogenic aerosols, the main objects of concern are sulfate aerosols, which are created through atmospheric chemical reactions involving sulfur dioxide, a gas produced by the combustion of fossil fuels. These tiny particles reflect incoming sunlight and, to a lesser degree, absorb infrared radiation. Perhaps more important, they also serve as condensation nuclei for clouds. When a cloud forms, water vapor does not form water droplets or ice crystals spontaneously but instead condenses onto preexisting aerosol particles. The number and size of these particles determine whether the water condenses into a few large droplets or many small ones, and this in turn strongly affects the amount of sunlight that clouds reflect and the amount of radiation they absorb.

It is thought that aerosols, in the aggregate, cool the planet because the increased reflection of sunlight to space—both directly by the aerosols themselves and through their effect on increasing the reflectivity of clouds—is believed to outweigh any increase in their greenhouse effect. Unlike the greenhouse gases, however, sulfate aerosols remain in the atmosphere for only a few weeks before they are washed out by rain and snow. Their abundance is proportional to their rate of production: as soon as their production decreases, concentrations of sulfate aerosols in the atmosphere drop. Since the late 1980s, improved technology and evermore stringent regulations, aided by the collapse of the USSR and the subsequent reduction and modernization of industrial output there, have diminished sulfate aerosol pollution in developed countries. On the other hand, sources of sulfate aerosols have been increasing in such rapidly developing countries as China and India, so the net aerosol content of the atmosphere may increase again.

Besides the uncertainties in the clouds and airborne particles that affect our climate system, there is another important source of uncertainty in attributing both past and future climate change to changes in solar radiation and atmospheric composition: our climate would change with time even if all of these factors were unchanging. This is because, like weather, climate varies all on its own. It is, on some level, a chaotic system.

The essential property of chaotic systems is that small differences tend to magnify rapidly. Think of two autumn leaves that have fallen next to each other in a turbulent brook. Imagine following them as they move downstream on their way to the sea: at first, they stay close to each other, but then the eddies in the stream gradually separate them. At some point, one of the leaves may get temporarily trapped in a whirlpool behind a rock while the other continues downstream. It is not hard to imagine that one of the leaves arrives at the mouth of the river days or weeks ahead of the other. It is also not hard to imagine that a mad scientist, having equipped our brook with fancy instruments for measuring the flow of water and devised a computer program that uses the measurements to predict where the leaves would go, would nevertheless find it almost impossible to pinpoint where the leaves would be even an hour after they started their journey.

Let's go back to the two leaves just after they have fallen in the brook and say that at this point they are 10 inches apart. Suppose that after 30 minutes they are 10 feet apart, and this distance increases with time. Now suppose that it were possible to rewind to the beginning but this time start the leaves only five inches apart. It would not be surprising if it took longer—say an hour—before they are once again 10 feet apart. Keep rewinding the experiment, each time decreasing the initial distance between the leaves. You might suppose that the time it takes to get 10 feet apart keeps increasing indefinitely. But for many physical systems (probably including brooks), this turns out not to be the case. As you keep decreasing the initial separation, the increases in the amount of time it takes for the leaves to be separated by 10 feet get successively smaller, so much so that there is a definite limit: no matter how close the leaves are when they hit the water, it will not take longer than, say, six hours for them to be 10 feet apart.

The same principle applies if, instead of having two leaves, we have a single leaf and a computer model of the leaf and the stream that carries it. Even if the computer model is perfect and we start off with a perfect representation of the state of the brook, any error—even an infinitesimal one—in the timing or position of the leaf when it begins its journey will lead to the forecast being off by at least 10 feet after six hours, and greater distances at longer times. *Prediction beyond a certain time is impossible.* It is important to understand that this limit to our ability to predict chaotic systems is a fundamental property of such systems; it is not possible, even in principle, to foresee the outcomes of such chaotic systems in detail beyond certain time limits.

Not all chaotic systems have this property of limited predictability, but our atmosphere and oceans, alas, almost certainly do. As a result, it is thought that the upper limit of the predictability of weather is around two weeks. (Our failure thus far to have reached this limit speaks to the imperfection of our models and our measurements.)

While the day-to-day variations of the weather are perhaps the most familiar examples of environmental chaos, variations at longer time scales can also behave chaotically. El Niño is thought to be chaotic in nature, making it difficult to predict more than a few months in advance. Other chaotic phenomena involving the oceans have even longer time scales.

On top of the natural, chaotic variability of weather and climate are changes brought about by variable "forcings," a term for agents of climate change that are not themselves strongly affected by climate. The most familiar of these is the march of the seasons, brought about by the tilt of the earth's axis, which itself is nearly independent of climate.[2] The effects of this particular forcing are not hard to separate from the background climate chaos: we can confidently predict that in New York, say, January will be colder than July, even though we cannot predict detailed weather there six months in advance. Other examples of natural climate forcing include variations in solar output and volcanic eruptions, which inject aerosols into the stratosphere and thereby cool the climate.

Some of this forcing is predictable on long time scales. For example, barring some catastrophic collision with a comet or asteroid, variations of the earth's orbit are predictable many millions of years into the future. On the other hand, volcanic activity is unpredictable. In any event, the climate we experience reflects a combination of "free" (unforced), chaotic variability, and changes brought about by external forcings, some of which, like volcanic eruptions, are themselves chaotic. Part of the recent forced climate variability has been brought about by human beings.

Distinguishing the forced response of the climate system from its chaotic natural variability requires a detailed understanding of the character of the latter, often referred to as "climate noise." Current estimates of this noise come largely from climate models run for a long time with constant forcing. These estimates suggest that the current global warming trend is clearly distinguishable from climate noise on time scales of around 30 years and longer. Just as a particular week in mid-spring may be colder than a particular week in late winter, there can be stretches as long as 30 years during which, owing to natural chaotic variability, the global mean temperature cools. Thus, for example, the lack of appreciable global warming over the first decade of the current millennium is, contrary to the claims of some, entirely consistent with the simultaneous occurrence of climate noise and greenhouse gas-induced warming. The record high temperatures of 2014, 2015, and 2016 put a predictable end to this "hiatus" (see figure 1).

Determining Humanity's Influence

How do we tell the difference between natural climate variations—both free and forced—and those that are caused by our own activities?

One way to tell the difference is to make use of the fact that the increase in greenhouse gases and sulfate aerosols dates back only to the Industrial Revolution of the nineteenth century: before that, the human influence is probably small. If we can estimate how climate changed before this time, we will have some idea of how the system varies naturally. Unfortunately, detailed measurements of climate did not themselves begin in earnest until the nineteenth century, but there are "proxies" for certain climate variables such as temperature. These proxies include the width and density of tree rings, the chemical composition of ocean and lake plankton, and the abundance and type of pollen.

Plotting the global mean temperature derived from actual measurements and from proxies going back a thousand years or more reveals that the recent upturn in global temperature is truly unprecedented: the graph of temperature with time shows a characteristic hockey-stick shape, with the business end of the stick representing the upswing of the last 50 years or so. The proxies are imperfect, however, and have large margins of error, so any hockey-stick trends of the past may be masked, but the recent upturn in global temperature still stands above even a liberal estimate of such errors.[1]

Another way to tell the difference is to simulate the climate of the last hundred years or so using computer models. Computer modeling of global climate is perhaps the most complex endeavor ever undertaken by humankind. A typical climate model consists of millions of lines of computer instructions designed to simulate an enormous range of physical phenomena, including the flow of the atmosphere and oceans; condensation and precipitation of water inside clouds; the transport of heat, water, and atmospheric constituents by turbulent convection currents; the transfer of solar and terrestrial radiation through the atmosphere, including its partial absorption and reflection by the surface, clouds, and the atmosphere itself; and vast numbers of other processes. There are by now a few dozen such models, but they are not entirely independent of one another, often sharing common pieces of computer code and common ancestors.

Although the equations representing the physical and chemical processes in the climate system are well known, they cannot be solved exactly. It is computationally impossible to keep track of every molecule of air and ocean, so to make the task viable, the two fluids must be divided up into manageable chunks. The smaller and more numerous these chunks, the more accurate the result, but with today's computers the smallest we can make these chunks in the atmosphere is around 50 miles in the horizontal and a few hundred yards in the vertical. We model the ocean using somewhat smaller chunks. The problem here is that many important processes happen at much smaller scales. For example, cumulus clouds in the atmosphere are critical for transferring heat and water upward and downward, but they are typically only a few miles across and so cannot be simulated by the climate models. Instead, their effects must be represented in terms of quantities such as wind speed, humidity, and air temperature that are averaged over the whole computational chunk in question. The representation of these important but unresolved processes is an art form known by the awkward term *parameterization*, and it involves numbers, or parameters, that must be tuned to get the parameterizations to work in an optimal way. Because of the need for such artifices, a typical climate model has many tunable parameters that one might think of as knobs on a large, highly complicated machine. This is one of many reasons that such models provide only approximations to reality. Changing the values of the parameters or the way the various processes are parameterized can change not only the climate simulated by the model, but also the sensitivity of the model's climate to, say, greenhouse gas increases.

How, then, can we go about tuning the parameters of a climate model so that it serves as a reasonable facsimile of reality? Here important lessons can be learned from our experience with those close cousins of climate models, weather-prediction models. These are almost as complicated and must also parameterize key physical processes, but because the atmosphere is measured in many places and quite frequently, we can test the model against reality several times per day and keep adjusting its parameters (that is, tuning it) until it performs as well as it can. In the process we come to understand the inherent accuracy of the model. But in the case of climate models, there are precious few tests. One obvious test is whether the model can replicate the current climate, including key aspects of its variability, such as weather systems and El Niño. It must also be able to simulate the seasons in a reasonable way: summers must not be too hot or winters too cold, for example.

Beyond a few simple checks such as these, however, there are not many ways to assess the models, and so projections of future climates must be regarded as uncertain. The amount of uncertainty in such projections can be estimated to some extent by comparing forecasts made by many different models, given their different parameterizations (and, very likely, different sets of coding errors). We operate under the expectation that the real climate will fall among the projections made with the various models—that the truth, in other words, will lie somewhere between the higher and lower estimates generated by the models. It is not inconceivable, though, that the actual solution will fall outside these limits.

While it is easy to stand on the sidelines and take shots at these models, they represent science's best effort to project the earth's climate over the next century or so. At the same time, the large range of possible outcomes is an objective quantification of the uncertainty that remains in this enterprise. Still, those who proclaim that the models are wrong or useless usually are taking advantage of science's imperfections to promote their own prejudices. Uncertainty is an intrinsic feature of prediction, and it works in both directions.

Figure 2 shows the results of two sets of computer simulations of the global average surface temperature during the twentieth century, using a particular climate model. In the first set, denoted by the dotted line and lighter shade of gray, only natural, time-varying forcings are applied. These consist of variable solar output and "dimming" owing to aerosols produced by known volcanic eruptions. The second set (dashed line and darker shade of gray) incorporates human influence on sulfate aerosols and greenhouse gases. Each set of simulations is run four times beginning with slightly different initial states, and the range of outcomes produced is denoted by the shading in the figure. This range reflects the random fluctuations of the climate produced by this model, while the bold curves show the average of the four ensemble members. The observed global average surface temperature is depicted by the black curve. The two sets of

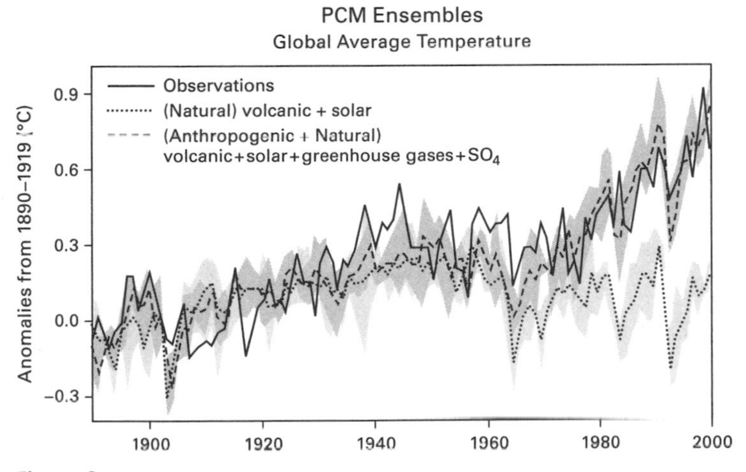

Figure 2
Showing the results of two sets of computer simulations of the global average surface temperature of the twentieth century using a climate model. In the first set, denoted by the dotted line and lighter shade of gray, only natural, time varying forcings are applied. The second set (dashed line and darker gray) adds in the man-made influences. In each set, the model is run four times beginning with slightly different initial states, and the range among the four ensemble members is denoted by the shading in the figure while the gray curves show the average of the four ensemble members. The observed global average surface temperature is depicted by the black curve.

simulations diverge during the 1970s and have no overlap at all today. The observed global temperature also starts to fall outside the envelope of the all-natural simulations in the 1970s.

This exercise has been repeated using many different climate models, with the same qualitative result: one cannot accurately simulate the evolution of the climate over the last 30 years without accounting for the human input of sulfate aerosols and greenhouse gases. This is one (but by no means the only) important reason that almost all climate scientists today believe that man's influence on climate has emerged from the background noise of natural variability. But the main reason remains the elementary physics that Arrhenius used to predict the global response to increasing greenhouse gases, long before the computer age.

The Consequences

Projections based on climate models suggest that the globe will continue to warm another 3–7°F over the next century. This is similar to the temperature change one could experience by moving, say, from Boston to Philadelphia. The warming of already hot regions—the tropics—is expected to be somewhat less than the global average warming, while the warming of cold regions like the arctic is projected to be more, a signal already clearly discernible in global temperature measurements. Nighttime temperatures are increasing more rapidly than daytime temperatures, and the temperature over continents is increasing faster than that over the oceans, consistent with both elementary theory and models.

Is this really so bad? With all the negative publicity about global warming, it is easy to overlook the benefits: it will take less energy to heat buildings, previously infertile lands of high latitudes may start producing crops, and there will be less suffering from debilitating cold snaps. Increased CO_2 should also make some crops grow faster, up to the point where their growth will be limited by the supply of nutrients. On the down side, there will be more frequent and more intense heat waves, air conditioning costs will rise, previously fertile areas in the subtropics may become barren, and blights may seriously affect both natural vegetation and crops.

So there will be winners and losers, but will the world really suffer in the net? Even if the changes we are bringing about are larger than the globe has experienced in the last few thousand years, they are smaller than the big natural swings between ice ages and interglacial periods, which the earth and indeed human beings survived.

The difference now, in a nutshell, is civilization, which developed during a period of exceptional climatic stability over the last 7,000 years. While our distant ancestors had to cope with a sea level rise of about 400 feet over a mere 8,000 years or so leading up to the development of civilization, human society has since become finely adapted to the current climate, so much so that a mere three-foot increase in sea level (small by the standards of geologically recent changes) would displace around 100 million people. Agriculture and animal husbandry are also tuned to the present climate, so that comparatively small shifts in precipitation and temperature can exert considerable pressure on governments and social systems whose failures to respond could lead to famine, disease, mass emigrations, political instability, and armed conflict.

Sea level rise is among the most serious potential consequences of global warming. During the peak of the last ice age, sea level was some 400 feet lower than it is today, since huge quantities of water were locked up in the great continental ice sheets. Today, almost all land ice is contained in the Greenland and Antarctic ice sheets. As polar regions warm up, it is possible that portions of these ice sheets will melt, increasing sea level. Highly detailed and accurate satellite-based measurements of the Greenland ice show that its thickness is actually increasing in the interior but thinning around the margins; in the net, there is significant loss of ice. Recent years have seen large increases in summer surface melting of the Greenland ice sheet. There are also patterns of increase and decrease in Antarctic ice, but it appears to be thinning on the whole. Meltwater from the surface of the Greenland ice sheet is making its way to the bottom of the ice, possibly allowing the ice to flow faster toward the sea. Our understanding of the physics of ice under pressure is poor, and it is thus difficult to predict how the ice will respond to warming. Were the entire Greenland ice cap to melt, sea level would increase by around 22 feet, flooding many coastal regions including much of southern Florida and lower Manhattan. Eleven of the fifteen largest cities in the world are located on coastal estuaries, and all would be affected. Thanks to variable patterns of winds and ocean temperature, sea level change is variable around the planet, so some regions could experience a greater increase than others. Moreover, sea level rise would exacerbate the effects of storm surges, which would occur more frequently and flood greater areas.

Other perils lie in wait. My colleagues and I have shown that hurricanes should become more intense and produce much more rain as the planet warms, and observations are beginning to show such trends. The 2005 Atlantic hurricane season was the most active in the 150 years on record, corresponding to record warmth of the tropical Atlantic. Katrina, which caused the largest storm surge in US history, cost more than $200 billion, and it claimed at least 1,200 lives. The 2017 Atlantic season was among the most destructive on record, causing well more than $300 billion in damages. The all-time record wind speed in a tropical cyclone was set by Typhoon Haiyan in 2013, only to be broken by Hurricane Patricia in 2015. Hurricane Sandy of 2012 had the largest diameter of any Atlantic storm on record, and 2017 Hurricane Irma set the world record for sustained Category 5 intensity, while Hurricane Harvey—in the same month—produced more rain than any hurricane in US history. Globally, tropical cyclones cause staggering misery and loss of life. Hurricane Mitch killed more than 10,000 people in Central America in 1998, and in 1970 a single storm took the lives of some 300,000 in Bangladesh. Substantial changes in hurricane activity cannot be written off as mere climate perturbations to which we will easily adjust.

Basic theory and models point toward another consequential result of a few degrees of warming: more floods and droughts. This happens because the amount of water vapor in the air rises exponentially with temperature: a 7°F increase in temperature increases the concentration of water vapor by 25 percent. One might suppose that the additional water ascending into clouds would produce a proportional increase in the amount of rain that falls out of them. But condensing water vapor heats the atmosphere, and in the grand scheme of things, this must be compensated by radiative heat loss. However, the amount of radiative heat loss increases only slowly with temperature, so that the total heating by condensation must increase slowly as well. Climate theory and models resolve this conundrum by making it rain harder in places that are already wet and at the same time increasing the intensity, duration, or geographical extent of droughts. Thus the twin perils of flood and drought both increase substantially in a warming world, and climate records show that this is happening.

The development of chronic food and water shortages in places that already have marginal climates, such as the Middle East, could easily lead to political destabilization. The US Department of Defense, in its February 2010 Quadrennial Defense

Review, succinctly summarizes these risks: Climate change could have significant geopolitical impacts around the world, contributing to poverty, environmental degradation, and the further weakening of fragile governments. Climate change will contribute to food and water scarcity, will increase the spread of disease, and may spur or exacerbate mass migration. In addition: While climate change alone does not cause conflict, it may act as an accelerant of instability or conflict, placing a burden to respond on civilian institutions and militaries around the world. Also, extreme weather events may lead to increased demands for defense support to civil authorities for humanitarian assistance or disaster response both within the United States and overseas.

It is particularly sobering to contemplate such outcomes in light of evidence that smaller, natural climate swings since the end of the last ice age debilitated and in some cases destroyed civilizations in such places as Mesopotamia, Central and South America, and the southwestern region of what is today the United States.

Beyond the direct impacts on climate, carbon dioxide emitted by human activities presents another danger: as atmospheric concentrations of carbon dioxide increase, roughly a quarter of the excess gas is absorbed by the ocean, increasing its acidity. Since the dawn of the Industrial Revolution, ocean water has become 30 percent more acidic. Among other potential problems, increasing acidity compromises the ability of a wide variety of marine organisms to form and maintain calcium carbonate shells. Such organisms are essential to the food chain: we threaten their welfare at our own peril. Declines in the rate of production of calcium carbonate also limit the oceans' ability to absorb our CO_2 emissions, leading to faster rates of increase in atmospheric concentrations.

The changes now occurring in the earth's climate and chemistry of the atmosphere and oceans are now so profound that in 2016 an expert group of geologists recommended to the International Geological Congress that the period since around 1950 be designated as a new geological epoch—the Anthropocene—in which the geological record is significantly influenced by human activity.

In pushing the climate and the oceans' geochemistry so hard and so fast, we should also be wary of our own collective ignorance of how the climate system works. Perhaps negative-feedback mechanisms that we have not contemplated or have underestimated will kick in, sparing us debilitating consequences. On the other hand, little-understood or unanticipated positive feedbacks might make matters worse than we expect. We are humbled by a sense of ignorance. For example, while we have come to understand that the great ice ages were caused by slow and predictable changes in the earth's orbit and rotation, we do not understand the sudden climate jumps revealed by the ice core record and are worried that similar jumps might be part of our future. We know as little about the consequences of our actions as Phaëthon did when he took the reins of his father's chariot.

Communicating Climate Science

Science proceeds by continually testing and discarding or refining hypotheses, a process greatly aided by the naturally skeptical disposition of scientists. Most of us are driven by a passion to understand nature, and that compels us to be dispassionate about pet ideas. Partisanship—whatever its source—is likely to be detected by our colleagues and hurt our credibility, the true stock of the trade. We share a faith—justified by experience—that at the end of the day, there are truths to be found, and those who cling, for emotional or ulterior reasons, to wrong ideas will be judged by history accordingly. Those who see the light early will be regarded as visionaries.

Although individual scientists are prone to the whole spectrum of human failings, the scientific endeavor as a whole is inherently conservative. Papers that scientists submit to respected journals must first be reviewed, usually anonymously, by their peers. This is only the first step in quality control and can result in recommendations for changes or in outright rejection. Once published, research is subject to critical review by other scientists, usually in proportion to the potential significance of the work. Are the results reproducible? Are the findings in accord with existing measurements? Do they make testable predictions? One way for a young and upcoming scientist to make a reputation is to disprove some accepted piece of the scientific canon; this is an important safeguard against the ubiquitous phenomenon of groupthink.

Thus science advances mostly by a series of faltering steps, two forward and one backward, as ideas are tested, disproved, and refined. On rare occasions, true revolutions occur, and large pieces of accepted wisdom have to be discarded or revised.

However well this process has served society, it does not usually mesh well with contemporary journalism, in which the drive to sell stories often elevates the provisional to the sensational. This is especially evident in the medical arena, where research papers reporting new findings on, say, the effect of diet on health are often seized upon and trumpeted by an eager press, only to be refuted by follow-up studies. In their quest to publish the drama of competing dogmas, the media largely ignore mainstream scientists whose hesitations make for dull copy.

Partly as an attempt to counteract this seesaw effect, climate scientists developed a way to communicate with the public and, at the same time, to compare notes and test one another's ideas. Called the Intergovernmental Panel on Climate Change (IPCC), it produces a detailed summary of the state of the science roughly every four years. The most recent report, the fifth in the series, was released in 2013 and 2014. Although far from perfect, the IPCC involves serious climate scientists from many countries and has done a superb job conveying up-to-date knowledge.

The IPCC reports are candid about what we know and where we think the uncertainties lie. Certain findings are not in dispute, not even among those generally skeptical of climate risk:

- Atmospheric concentrations of key greenhouse gases—carbon dioxide, methane, ozone, and nitrous oxide—are increasing, driven by the burning of fossil fuels and biomass. Carbon dioxide has increased from its pre-industrial level of about 280 parts per million to about 405 parts per million today, an increase of about 45 percent. From ice-core records, it is evident that present levels of CO_2 are higher than they have been in at least the last 800,000 years.

- Concentrations of certain aerosols have increased owing to industrial activity.
- The earth's average surface temperature has increased by about 1.5°F in the past century, with most of the increases occurring from about 1920 to 1950, and again beginning around 1975. The year 2016 was the warmest in the instrumental record so far, closely followed by 2015 and 2014.
- Global average sea level has risen by about 8.4 inches since 1880. A little more than an inch of this rise occurred during the past decade. Sea level rise is not, however, globally uniform and is rising faster in some places and slower in others.
- The annual mean geographical extent of arctic sea ice has decreased by 15–20 percent since satellite measurements began in 1978.
- The acidity of ocean water has increased by about 30 percent since the beginning of the industrial era.
- The global mean temperature is now greater than at any time in at least the last 500 years.
- Most of the global mean temperature variability over the past century was caused by variability of solar output, major volcanic eruptions, and anthropogenic sulfate aerosols and greenhouse gases.
- The dramatic rise in global mean temperature in the past 30 years is attributable primarily to increasing greenhouse-gas concentrations and a leveling off or slight decline in sulfate aerosols.

There is a second category of findings that most—but not all—climate scientists agree with:

• Unless measures are taken to reduce greenhouse gas emissions, global mean temperature will continue to increase. The rise will be between 2.5 and 9°F over the next century, depending on uncertainties and how much greenhouse gas is produced.

• As a result of thermal expansion of seawater and melting of polar ice caps, sea level will increase 7 to 23 inches over the next century, though the increase could be even greater if large continental ice sheets become unstable.

• Rainfall will continue to become concentrated in heavier but less frequent events.

• The incidence, intensity, and duration of both floods and drought will increase.

• The frequency of the most intense hurricanes is likely to increase noticeably, and hurricanes will produce more rain and thus freshwater flooding. The combination of increasing intensity and rising sea level foretells increased coastal flooding by storm surges.

• The acidity of ocean water will continue increasing.

Even if we could be certain about the increases of greenhouse gases and aerosols, estimating their net effect on humanity is an enormously complex undertaking, pitting uncertain estimates of costs and benefits against the costs of curtailing greenhouse gas emissions. We are by no means sure of what changes are in store, and we must be wary of climate surprises. Even if we were to take the extreme view that the projected climate changes would be mostly beneficial, we might be inclined to make sacrifices as an insurance policy against potentially harmful surprises.

Our Options

Global climate change presents us with unprecedented challenges. Since science can do no more than estimate a broad envelope of possible outcomes, from the benign to the catastrophic, society must treat the problem as one of risk assessment and management. At one extreme, we could elect to do nothing and gamble on a benign outcome. But if we are wrong we will saddle our grandchildren and their descendants with enormous problems. At the other, we could make serious economic and other tangible sacrifices that might prove unnecessary. Unfortunately, waiting much longer to see which way things go is not a viable option, as it takes thousands of years for CO_2 levels to return to normal once emissions cease. By the time the consequences of climate change become unequivocally clear, it will almost certainly be too late to do much about the potentially massive problems.

Scientists, engineers, and economists can do no more than formulate options for dealing with the risks. It is up to society as a whole to decide what combination of options to deploy. This is a terrifically difficult decision because the costs may be high and those paying them are not likely to be serious beneficiaries of their own actions. Indeed, there are few, if any, historical examples of civilizations consciously making sacrifices on behalf of descendants two or more generations removed.

Options for dealing with climate change fall into three broad categories: curtailing the emissions of greenhouse gases (mitigation), learning to live with the consequences (adaptation), and engineering our way around the problems that greenhouse gases produce (geoengineering).

Of these three options, mitigation has the most straightforward effect on climate because it attacks the source of the problem. Although they can be costly, some aspects of mitigation might be worth undertaking anyway. For example, consumers might spend extra money on a high-efficiency car if the excess cost is paid back in fuel cost savings over a few years. Similarly, the costs of constructing or retrofitting buildings to conserve energy might also be paid back in a short time. Such conservation measures would not only help reduce emissions but would also prove economically beneficial for consumers.

But given the expected growth in the economies of developing nations such as China and India, conservation alone cannot begin to reduce greenhouse gas emissions to safe levels. Experience has shown unequivocally that rapid economic growth can be achieved only with large increases in per capita energy consumption. The alleviation of the wrenching poverty of poor nations is, of course, a highly desirable goal, but it also appears to be a necessary condition for the reduction of population increase, which is a key driver of energy growth. Thus the global problems of climate, energy, poverty and population are inextricably linked.

Fortunately, the means of de-carbonizing energy are at hand. The growth in solar and wind power in recent decades has been truly impressive, and the price of these energy sources has fallen as demand increases and the technology improves. Even so, solar, wind, and hydroelectric provide only 8 percent of global electrical power today, and most energy experts believe that the inherent intermittency of these sources will limit their market penetration to 30–40 percent, barring true breakthroughs in energy storage and transmission technologies.

Nuclear fission provides about 11 percent of global electrical energy but today relies entirely on light water reactors that operate at high pressure and produce radioactive waste. Even so, nuclear fission is far and away the safest form of energy humankind has ever produced, with mortality per kilowatt hour generated less than that of any other energy source, including solar and wind. While much is made of events such as that at the Fukushima facility in Japan, petrochemical accidents brought about by the earthquake and tsunami killed many while no deaths resulted from Fukushima's release of radioactive material. Indeed, it is estimated that nuclear fission has saved about 1.8 million lives by displacing fossil fuels, whose combustion is the source of numerous health problems.

Yet nuclear technology has advanced significantly since light water reactors were introduced more than a half-century ago. Advanced reactors operate at ambient pressure and are passively safe, so they are inherently incapable of melting down. They burn fuel far more efficiently, resulting in greater power production per unit input of fuel, and much less radioactive waste. They are far more environmentally benign than solar or wind, requiring much less land, and many of the new designs require very little water for cooling.

Actual experience in countries such as Sweden and France shows that fission power can be ramped up to supply a large fraction of electrical energy in just 15 years.[1] What is now lacking more than anything else is political will.

Another mitigation strategy is to reduce the effect of emissions by capturing and storing their greenhouse gas components. Such technology exists today but is currently estimated to increase the cost of energy produced by 20–90 percent. There is some hope that technological developments might bring these costs down. Capturing carbon at its industrial source is perhaps the best of all solutions if it can be done economically, because fossil fuels are so abundant and affordable and because extensive infrastructure already exists for producing and distributing them. It is also possible to capture CO_2 directly from the atmosphere, but this is currently much more expensive because atmospheric concentrations of the gas are far lower than those at the emissions sources.

The implementation of carbon capture and storage, higher efficiency vehicles and buildings, and other mitigation measures could be accelerated by a variety of governmental actions, such as carbon taxes, cap-and-trade policies, and subsidies for carbon-free energy. But here the problem of reducing emissions is most inextricably bound with partisan politics. At first sight there appears to be little common ground between those viscerally opposed to any government action that increases the cost of energy and those determined to reduce emissions even if doing so would tangibly hurt the economy. But on closer inspection it becomes apparent that our failure to migrate away from fossil fuels owes as much to government interference in free markets as it does to excesses of laissez-faire capitalism. The US government provides billions of dollars in annual subsidies to the coal, oil, and natural gas industries. Reducing or eliminating these subsidies would free up markets, make alternative energy sources more competitive, and motivate energy companies to develop cleaner alternatives. Free-market principles don't allow businesses to pass part of their costs on to other enterprises, yet the health costs of coal mining and combustion alone are estimated at $65–185 billion annually.[2] Currently these costs must be paid by private health insurance ratepayers and by the taxpayer, through government-funded health insurance. In a true free-enterprise system, all businesses would cover their external as well as internal costs. Insisting that the energy industry do so would, in addition, naturally favor cleaner alternatives.

Whereas the costs of mitigation fall mostly on the largest emitters of greenhouse gases, the costs of adaptation are more broadly distributed over the world. For example, the low-lying Pacific island nation of Kiribati, with a population of just over 110,000, is threatened by rising sea levels, and the current government plans to begin moving the entire population to Fiji in 2020.[3] At the other extreme, countries such as Russia and Canada might profit from a warmer climate. But most nations will need to adapt to climate change, entailing measures ranging from crop substitutions to beefing up sea walls and levees and planning for shifting demands for and supplies of water and food.

A key but complex issue is the relative costs and benefits of adaptation and mitigation, all of which must be estimated in an environment of considerable uncertainty. An optimal strategy will no doubt involve doing some of both.

The third approach, geoengineering, seeks to actively counter greenhouse gas–induced warming. Proposals aimed at cooling the earth focus primarily on managing the net amount of solar radiation the planet absorbs by increasing the reflectivity (albedo) of the surface and the atmosphere. A popular technique involves injecting modest amounts of sulfur into the stratosphere, resulting in the formation of sulfate aerosols that reflect sunlight and thereby cool the climate system. The technology to do this exists today, and the cost of doing so is small enough that a small nation or even a wealthy individual could pull it off.

But there are many technical, legal, and political problems with solar radiation management. On the technical side, cooling the mean surface temperature back to some desired point (say, enough to prevent damaging sea-level rise) while leaving atmospheric concentrations of CO_2 unabated would not necessarily repair other important aspects of the climate system. In particular, canceling a long-wave radiative effect (greenhouse gas warming) with a short-wave fix (reflecting solar radiation) does not necessarily restore variables other than temperature. For example, bringing the temperature back to some desired level would almost certainly result in a reduction of global precipitation. Moreover, engineering solar radiation does nothing to address the CO_2-induced acidification of the oceans that may prove to be among the most serious consequences of greenhouse gas emissions. Furthermore, any entity, whether an individual or a nation, that undertook geoengineering would do so within a largely undeveloped legal framework, leaving it exposed to legal or even military action. For all these reasons, most of those whose work focuses on geoengineering regard it as an option to be developed and then kept in our collective back pocket, to be used only if the effects of climate change become catastrophic.

The Politics Surrounding Global Climate Change

Especially in the United States, the political debate about global climate change became polarized along the conservative-liberal axis some decades ago.

Although we take this for granted now, it is not obvious why the chips fell the way they did. The Republican Party has a respectable track record in protecting the environment, from Abraham Lincoln's deeding of Yosemite Valley to California, to Richard Nixon's establishment of the Environmental Protection Agency and signing of the Clean Air Act, to Ronald Reagan's strong advocacy of the Montreal Protocol, which sought to protect the ozone layer, and George H. W. Bush's support of the United Nations Framework Convention on Climate Change and the historic 1990 amendments to the Clean Air Act. One can easily imagine conservatives embracing climate policies that are in harmony with other actions they might like to see. Conservatives have usually been strong supporters of nuclear power, and few can be happy about our partial dependence on foreign oil. Neither can conservatives be comfortable with taxpayer subsidies of fossil fuels or the absorption of coal externalities into the price of health care. Combating climate change offers business opportunities that conservatives should embrace. The United States is renowned for its technological innovation and should be at an advantage in making money from any global change in energy-producing technology: consider the prospect of selling high-efficiency vehicles, electrical generation, and carbon capture and sequestration technology to China's rapidly expanding economy. But none of this has happened. Indeed, conventional wisdom among Republicans has shifted in recent years from questioning the economics of climate policy proposals (a comparatively centrist stance) to an increasingly strident skepticism about climate science itself. The US withdrawal from the Paris Agreement in 2017 effectively ceded a potential $6 trillion market in carbon-free energy to China, an action that conservatives will soon come to see as economic folly of the first order.

Contradictions abound on the political left as well. A meaningful reduction in greenhouse gas emissions will require a significant shift in the means of producing energy, as well as conservation measures. But such alternatives as nuclear power are viewed with deep ambivalence by the left, and only a few environmentalists have begun to rethink their visceral opposition to it. Had it not been for green opposition, the United States today might derive most of its electricity from nuclear sources, as France does. Thus environmentalists must accept some measure of responsibility for today's most critical environmental problem. Indeed, by focusing on solar and wind power sources—whose intermittency prevent them from meeting more than a fraction of our energy needs—the environmental movement is engaged in unproductive theater that detracts from serious debate about energy.

It is important to note here that by not developing and deploying new nuclear technology, developed countries are, ironically, putting the world at greater risk of nuclear terrorism. Russia and China are currently building reactors based on old light water technology and selling them or leasing them abroad, increasing the proliferation of vulnerable fissile material. Nations such as the United States cannot stop this by fiat but could out-compete Russia and China by developing safer, cheaper, and thus more attractive alternatives to sell to their clients.

The important lesson here is that the global issues of climate, poverty, energy, national security, and national prosperity are bound together and can no longer be contemplated separately.

Climate research has been a victim of a disturbing phenomenon: the use of advanced marketing techniques to discredit scientific findings that may lead to consumer and regulatory behavior unfavorable to certain business interests. An early example was the highly successful campaign waged by the tobacco industry to persuade consumers that there is little or no health risk from smoking. Such campaigns capitalize on scientists' equivocal disposition and journalists' fondness for controversy to sow doubt on a massive scale.[1] It did not prove difficult to persuade those addicted to nicotine that the alarms being raised were manufactured by corrupt and incompetent scientists in league with socialist-leaning do-gooders. The campaign delayed by roughly 30 years the public response to scientific evidence linking smoking to cancer, evidence that had become strong by the early 1950s. In so doing, the pro-smoking campaign cost many millions of lives.

The campaign to discredit climate science dates at least as far back as the early 1990s. For example, in 1991 the Western Fuels Association established the Information Council for the Environment (ICE) to "demonstrate that a consumer-based media awareness program can positively change the opinions of a selected population regarding the validity of global warming."[2] The Council planned an ad campaign that would "directly attack the proponents of global warming by relating irrefutable evidence to the contrary, delivered by a believable spokesperson" and would "attack proponents through comparison of global warming to historical or mythical instances of gloom and doom." The campaign specifically targeted older, less-educated males and younger, lower-income women. Market research conducted under ICE's auspices suggested that the public trusts technical sources more than it does politicians or industry representatives, so ICE sought scientists who were skeptical of the idea that climate change posed significant risk. Within any scientific endeavor there are always mavericks, and they play an important role in the constant introspection that helps ward off groupthink and other perils to progress. It is not difficult for extra-scientific organizations to amplify these maverick voices so as to create the illusion of serious controversy. This tactic, greatly aided by journalists' attraction to controversy, often masquerading as balance, has been particularly successful in denigrating mainstream climate science.

Other components of a successful campaign to cast doubt on scientific findings include conflating uncertainty with ignorance, associating scientists with extremists and otherwise impugning their motives, and planting the romantic idea that mavericks are often right while scientific organizations, such as the National Academy of Sciences, are often wrong. Most people, when it comes to their personal health, would never ignore the advice of 97 doctors in favor of three. But through the wondrous alchemy of marketing, it is possible to get some people to do just that in the realm of climate science.

There are other obstacles to taking a sensible approach to the climate problem. We have had precious few representatives in Congress with a background or interest in science, and some of the others actively display contempt for the subject. As long as we continue to elect and appoint scientific illiterates such as James Inhofe and Scott Pruitt, who believe global warming to be a hoax, we will be discouraged from engaging in intelligent debate at the policy level.

On the bright side, the governments of many countries continue to fund climate research, and many of the critical uncertainties about climate change are slowly being whittled down. The extremists are being exposed and relegated to the sidelines, and if the media stop amplifying their views, their political counterparts will have nothing left to stand on. When this happens, we can get down to the serious business of tackling the most complex and perhaps the most consequential problem ever confronted by humankind. In so doing, we will improve both our economy and the environment by developing clean energy sources that will carry us forward for many generations.

Notes

Chapter 2: Greenhouse Physics

1. A cautionary note: the greenhouse metaphor is flawed because actual greenhouses work mostly by preventing convection currents from carrying away heat absorbed from sunlight, not by preventing heat from radiating away from a surface.

Chapter 3: Why the Climate Problem Is Difficult

1. Indeed, it was well understood by the end of the nineteenth century. In 1897 the Swedish chemist Svante Arrhenius published a paper estimating that doubling atmospheric CO_2 would increase earth's surface temperature by 9–14°F.
2. Formation and melting of ice sheets and even the movement of the oceans and atmosphere cause variations in the earth's rotation and orbit, but these are too small to really affect climate.

Chapter 4: Determining Humanity's Influence

1. The hockey-stick shape of the temperature curve over the past few thousand years has become something of an icon of anthropogenic climate change and has been strongly criticized from many quarters. But repeated analyses of the underlying proxy data have verified the basic

shape of the temperature curve and have shown the last century of warming to be anomalously pronounced.

Chapter 7: Our Options

1. Raymond Pierrehumbert, "How to Decarbonize? Look to Sweden," *Bulletin of the Atomic Scientists*, 72, no. 2 (2016): 105–111.
2. Paul R. Epstein, Jonathan J. Buonocore, Kevin Eckerle, et al., "Full Cost Accounting for the Life Cycle of Coal," *Annals of the NY Academy of Science* 1219 (2011): 73–98.
3. Alex Pashley, "Kiribati President: Climate-Induced Migration Is 5 Years Away," *Climate Home News*, February 18, 2016. http://www.climatechangenews.com/2016/02/18/kiribati-president-climate-induced-migration-is-5-years-away (accessed April 26, 2018).

Chapter 8: The Politics Surrounding Global Climate Change

1. For example, the Brown and Williamson Tobacco Corporation's internal document, "Smoking and Health Proposal" (1969), states, "Doubt is our product since it is the best means of competing with the 'body of fact' that exists in the mind of the general public. It is also the means of establishing a controversy." See http://legacy.library.ucsf.edu/tid/nvs40f00 (accessed July 18, 2012).
2. International Council for the Environment, "ICE Mission Statement," 1991. Archives of the American Meteorological Society, Boston, MA.

Further Reading

Alley, R. B. *The Two-Mile Time Machine: Ice Cores, Abrupt Climate Change, and Our Future*. Princeton, NJ: Princeton University Press, 2014. 248 pp. An engaging look into what we have learned about earth's climate history from the analysis of ice cores from Greenland and Antarctica, showing that climate can change very rapidly.

Archer, D., and R. Pierrehumbert, eds. *The Warming Papers*. Chichester, UK: Wiley-Blackwell, 2013. 432 pp. A wonderful collection of the most important and influential papers in the history of climate science.

Archer, D. *The Long Thaw: How Humans Are Changing the Next 100,000 Years of Earth's Climate*. Princeton, NJ: Princeton University Press, 2016. 200 pp. A compelling overview of climate science and its implications for the long-term future of climate and humanity.

Diamond, J. *Collapse: How Societies Choose or Fail to Succeed*. New York: Penguin Books, 2011. 608 pp. A fascinating account of how past civilizations coped with, or failed to cope with, ecological crises, set in the context of our own time and the current climate challenge.

Emanuel, K. A. "Climate Science and Climate Risk: A Primer," 2016. ftp://texmex.mit.edu/pub/emanuel/PAPERS/Climate_Primer.pdf. A comprehensive but brief overview of evidence for human-induced climate change.

Houghton, J. *Global Warming: The Complete Briefing*. Cambridge: Cambridge University Press, 2015. 396 pp. A companion text to the IPCC report, this presents a complete overview of climate science, including climate physics and past climates, and also a useful description of climate models and their projections.

Houser, T., and S. Hsiang. *Economic Risks of Climate Change: An American Prospectus*. New York: Columbia University Press, 2015. 384 pp. A detailed economic analysis of the monetary costs of climate change in the United States, across many affected sectors, from agriculture to crime.

Intergovernmental Panel on Climate Change. "Climate Change 2013, the Physical Science Basis," 2013. https://www.ipcc.ch/report/ar5/wg1. A comprehensive summary of theoretical, modeling, and observational evidence of climate change.

Keith, D. *A Case for Climate Engineering*. Boston: Boston Review Books; Cambridge, MA: MIT Press, 2013. 220 pp. A good summary of proposed techniques for mitigating climate change through geoengineering, and a compelling case for at least considering the use of these techniques for bettering our condition.

Kolbert, E. *The Sixth Extinction: An Unnatural History*. New York: Picador, 2015. 336 pp. A highly engaging overview of past global extinctions and evidence that we are entering into a period of another such extinction.

Nordhaus, W. D. *The Climate Casino: Risk, Uncertainty, and Economics for a Warming World*. New Haven: Yale University Press, 2015. 392 pp. A world-leading economist explains why acting now will end up costing less than waiting.

Oreskes, M., and E. M. Conway. *Merchants of Doubt: How a Handful of Scientists Obscured the Truth on Issues from Tobacco Smoke to Global Warming*. New York: Bloomsbury Press, 2011. 368 pp. A riveting account of how corporations and renegade scientists use advanced marketing techniques to promulgate harmful lies.

Partanen, R., and J. M. Korhonen. *Climate Gamble: Is Anti-Nuclear Activism Endangering Our Future?* Helsinki: Viestintätoimisto Cre8 Oy, 2017. 132 pp. Documents the campaign of fear of nuclear power by environmental activists, and how this is impeding the migration to carbon-free energy.

Pierrehumbert, R. T. *Principles of Planetary Climate*. Cambridge: Cambridge University Press, 2011. 674 pp. A comprehensive text on climate science aimed at beginning graduate students. Very thorough, this is the best textbook available at this level.

Sobel, A. *Storm Surge: Hurricane Sandy, Our Changing Climate, and Extreme Weather of the Past and Future*. New York: Harper Wave, 2014. 336 pp. Adam Sobel presents an entertaining but sobering account of Hurricane Sandy in the context of global climate change.